OUR ONE ZANY UNIVERSE

Naira R. Matevosyan, Richard Matevosyan

ISBN: 978-1506141312. - Createspace, Inc; Seattle, WA

1 - 38

One starry night in September, father jaguar and the cub took a tour of the space beyond *Milky Way* and *Andromeda,* on the path to the *Bing Bang* theory.

"If you were asked to define the logic of universe in a word, would it be the *gravity?*" meowed the cub.

"It is part of the logic. We have to delve into some results of the gravitational theory to understand the universe. Gravity sets up rules, standards, relationships, orbiting paths, and its power even bends the light. For example, the objects we are looking at, might not actually be where they appear to be," rambled the jaguar.

"Does it also govern the death?" purred the cub.

"I see what you mean: the black whole or the monstrous *Gorgonzola...*" thundered

father jaguar. "Loosely speaking, a black hole is a region of space that has so much mass concentrated in it, and therefore it exhibits such a strong gravitational pull, that no imminent particle can escape from it. Even the light itself cannot escape the black hole. That coldest spot is the ultimate *cemetery* of everything created and known in the universe. The swirling gases around a black hole turn it into an electrical generator, making it spout jets of electricity billions of kilometers out into space."

"I see a conflict in your interpretation - between the theory of density and the law of diffusion or osmosis," mewled baby jaguar.

"To resolve that conflict you first need to define *'theory'* or *'physics.'* In simpler terms, physics is the theory of nature or properties of the matter, energy, relationships, or phenomenon. As a vacuum, universe rules out most of the canons applicable on Earth. Make a guess: what would happen to an ordinary glass of water placed in vacuum, in outer space?"

"I've no idea. Perhaps it would evaporate?"

"Nope. It would first boil, then freeze to solid," roared the jaguar.

"How come?"

"At sea level, pure water boils at 100°C. That's the commonly known boiling point. On the top of Mount Everest however, where the air pressure is too low, water boils at roughly 65°C. Thus, it makes very challenging to make a cup of tea on top of Everest. In Tibet for example, the monks add milk and sugar before they boil the tea, so these 'impurities' would increase the boiling point enough to make the tea more enjoyable!

In outer space, in the vacuum, water would begin to boil immediately not because of the temperature, but because of the lack of pressure to hold the molecules together."

"What about freezing?" meowed the cub.

"Let's simulate our experience by placing this ordinary glass of water in the

vacuum chamber. We create vacuum by pumping out all the air inside the chamber. As the air is pumped out completely, the pressure decreases and.... the water begins to boil - even though it's not hot. After a few minutes, the water will stop boiling!"

"Why?" meowed the cub.

"Because the change of properties from a liquid to gas took a lot of energy - in form of heat - away from the water, causing its temperature to drop until it froze."

"This explains, why most of the celestial bodies, like comets or asteroids, are mainly formed of ice and dust."

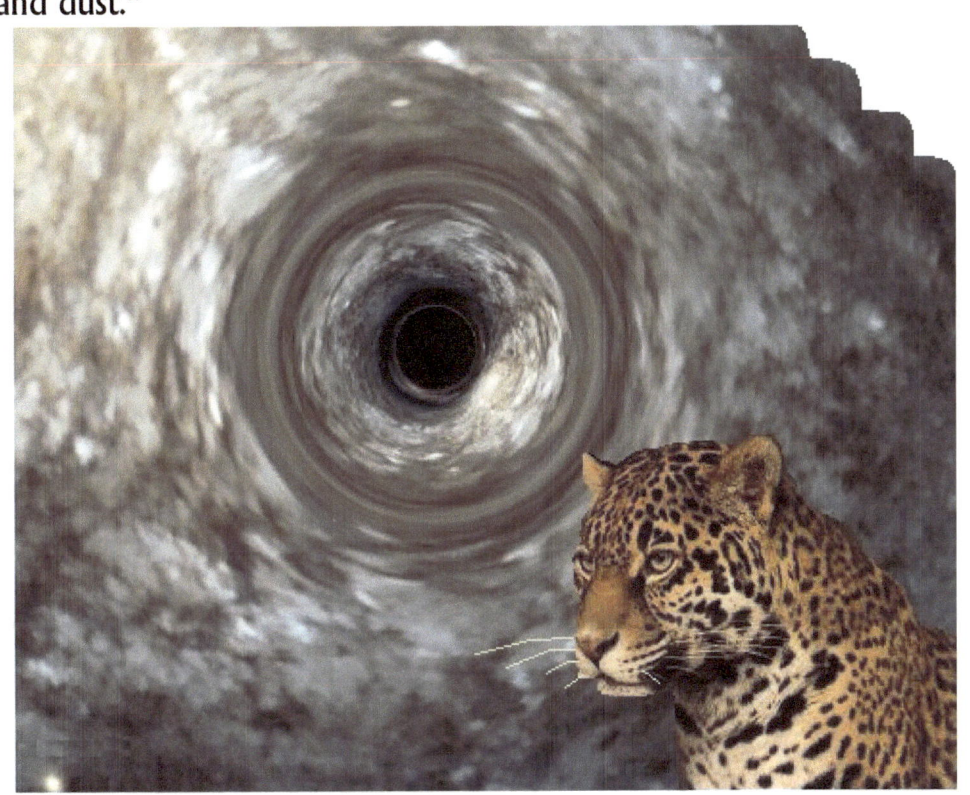

"Partly so, as there are also other factors involved."

"If I got it right, massive objects distort space and time with their gravity, so that the usual rules of geometry don't apply anymore."

"Quite right.

Near the black hole, this distortion of space is extremely severe. It causes black holes to have very ill properties. In particular, arguably though, a black hole has something called an *'event horizon.'* This is a spherical surface that marks its boundary. You can pass in through the horizon, but you can't get back out. In fact, once you've crossed the horizon, you're doomed to move inexorably closer and closer to the *'singularity'* at the center of the black hole," rambled the jaguar.

"Is there a mathematical explanation for this *burial?*" asked the cub.

"The idea of a *'mass concentration so dense that even light would be trapped'* goes all the way back to *Einstein's* general relativity theory followed by its mathematical solution suggested by *Karl Schwarzschild.* There are two relativity theories: special, and general. In *general relativity*, gravity is a manifestation of the curvature of space-time:

$$G_{\mu\nu} + \Lambda g_{\mu\nu} = \frac{8\pi G}{c^4} T_{\mu\nu}$$

Under this principle of equivalence, the states of accelerated motion and at the rest in a gravitational field (for example when standing on the surface of the Earth) are physically identical. The upshot is that *free fall is an inertial motion*: in classical mechanic, an object in free fall is effected by no force but its gravity. This is incompatible with *special relativity theory,* according to which inertly moving objects cannot accelerate with respect to each other, but objects in  free fall can do so. To resolve this conflict Einstein first proposed that *space-time is curved.* In 1915, he devised the Einstein field equations which relate the curvature of space-time with the mass, energy, and momentum within it.

Much later in the 1930's, Oppenheimer (the one who also ran the Manhattan Project), Volkoff, and Snyder seriously revered to a possibility that such ill objects might actually exist in the universe. These luminary-minds suggested that when a sufficiently massive star runs out of the fuel, it is then unable to support itself against its own gravitational pull, and it should collapse into a black hole."

"And that brings to the *string theory*," replied the cub.

"Correct. One of the longest and outstanding mysteries in physics is how gravity is related to the other fundamental forces, such as electromagnetism. One theory proposed in 1919, suggested that if an extra dimension is added to the universe, gravity still exists in the first four dimensions (3D + time = 4D), but the way this four dimensional space curves over the extra fifth dimension (5D) naturally produces the other fundamental forces. However, we cannot see or detect this fifth dimension, so it was proposed that the extra dimension was curled up, and hence became invisible to us. This theory was what ultimately led to the *string theory*, or *Kaluza-Klein (KK) theory*, and is still included at the heart of most string theory analysis.

In 1926, Oskar Klein gave Kaluza's classical 5D theory a quantum interpretation to accord with the new discoveries of Heisenberg and Schroedinger. Klein introduced the hypothesis that the fifth dimension was curled up and microscopic, to explain the cylinder condition. Klein also calculated a scale for the 5D based on the quantum of charge. But it wasn't until the 1940s that the classical theory was completed and the full field equations, including the scalar field, were obtained by three independent research groups: *Thiry* – working in France; *Jordan, Ludwig,* and *Müller -* in Germany; and *Scherrer -* working alone in Switzerland. [1-4]

Precisely though, the 5-dimensional metric has 15 components. Ten are identified with the 4D spacetime metric, four with the electromagnetic vector potential, and one with an unidentified scalar field, sometimes called the *'radion'* or the *'dilaton.'* Correspondingly, the 5-dimensional Einstein equations yield the 4-dimensional Einstein field equations, the Maxwell equations for the electromagnetic field, and an equation for the scalar field. *Kaluza* also introduced the hypothesis known as the *'cylinder condition,'* that no component of the 5D metric depends on the fifth dimension. Without this assumption, the field equations of 5-dimensional relativity are enormously more complex. Standard 4-dimensional physics seems to manifest the cylinder condition. *Kaluza* also set the scalar field equal to a constant, in which case standard general relativity and electrodynamics are recovered identically.

Later in 1971, German astrophysicists *Dehnen* and *Obregón* suggested an *exact solution of cosmological equations for the scalar-tensor theory,* showing that *a power law exists between the gravitational constant and the radius of curvature of the universe. For the space of negative curvature no solution is possible. On the contrary, for a closed space, the gravitational constant and the radius of curvature linearly increase with respect to the age of the universe.* [5]

"This solution has no analogy in Einstein's theory with vanishing cosmological constant, even when the deviations from Einstein's values of the solar relativistic effects are small," noticed the cub.

"Good point!" roared the jaguar. "One particular variant of *KK theory* is space-time-matter theory, or *induced matter theory,* suggesting that solutions to the equation

$$\tilde{R}_{ab} = 0$$

may be re-expressed so that in four dimensions, these solutions satisfy Einstein's equations

$$G_{\mu\nu} = 8\pi T_{\mu\nu}$$

with precise form of the $T_{\mu\nu}$ following from the Ricci-flat condition on the 5D space. In other words, the cylinder condition of the previous development is dropped, and the stress-energy now comes from the derivatives of the 5D metric with respect to the fifth coordinate. Since the energy–momentum tensor is normally understood to be due to concentrations of matter in 4D space, the above result is interpreted as *four-dimensional matter is induced from geometry in five-dimensional space.*" [6]

"It is my view," continued the cub, "that since in our positive spacetime the extra dimension is too small and curled up on itself, only tiny particles can move along it and they ultimately just end up where they started."

"True. However, one object that becomes much more complex in five dimensional spacetime is the *black hole*. When extended to 5D, it becomes a *'black string.'* Unlike a regular 4D black hole, it is unstable. This black string will destabilize into a whole string of black holes, connected by further black strings, until the black strings are pinched off entirely and leave the set of black holes. These multiple 4D black holes then combine into one larger black hole. The most interesting thing about this, is that using current models the final black hole is a *naked singularity.* That said, it has no event horizon surrounding it. This violates the *Cosmic Censorship Hypothesis*, which suggests that all singularities must be surrounded by an event horizon, in order to avoid the time-travel effects that are believed to happen near a singularity from changing the history of the entire universe, as they can never escape from behind an event horizon."

"I can think of the horizon as the place where the

escape velocity equals the velocity of light. Outside of the horizon, the escape velocity is less than the speed of light, so if you fire your rockets hard enough, you can give yourself enough energy to get away. But if you find yourself inside the horizon, then no matter how powerful your rockets are, you can't escape."

"Indeed. The horizon has very strange geometrical properties. To an observer who is sitting still somewhere far away from the black hole, the horizon seems to be a nice, static, steady spherical surface. But once you get close to it, you realize that horizon has a very large velocity. In fact, it is moving outward at the speed of light! That explains why it is easy to cross the horizon in the inward direction, but impossible to get back out. Since the horizon is moving out at the speed of light, in order to escape back across it you would have to travel faster than light. You can't go faster than light, and so you can't escape from the black hole!" thundered father jaguar.

"Here's your problem. When you say *'faster,'* your *terrestrial* definition of speed, as in Sir Newton's law (S = D/T), works only in our great planet Earth," noticed the cub. "If *Speed = Distance/Time*, the motion of a moving object can be explained either by Newton's Law and vector principles, or by means of the *Work-Energy Theorem*. This is true also in explaining the parabolic or circular motions of a projectile. Suppose, you are driving on an *autobahn* and are stopped by the Bavarian Police, after the steering wheel turned in such a manner that your *Opel Ascona* followed the path of a perfect circle with a constant radius. And suppose that as you drove, your speedometer maintained a constant reading of 40 m/sec. In such a situation as this, the motion of your car could be

15 - 38

described as experiencing *uniform circular motion*. It is the motion of an object in a circle with a constant or uniform speed. An object moving in *uniform circular motion* would cover the same linear distance in each second of time. When moving in a circle, an object traverses a distance around the perimeter of the circle."

"So?"

"So, if your car were to move in a circle with a constant speed of 40 m/s, then the car would travel 40 meters along the perimeter of the circle in each second of time. The distance of one complete cycle around the perimeter of a circle is the circumference. With a uniform speed of 40 m/s, a car could make a complete cycle around a circle that had a circumference of 40 meters. At this uniform speed, each cycle around the 40-m circumference circle would require 1 second. At 40 m/s, a circle with a circumference of 160 meters could be made in 4 seconds; and at this uniform speed, every cycle around the 160-m circumference of the circle would take the same time period of 4 seconds. This relationship between the circumference of a circle, the time to orbit one cycle, and the speed of the object is merely an extension of the average speed equation stated in

$$\text{Average Speed} = \text{Distance/Time} = \text{Circumference/Time}.$$

The circumference of any circle can be computed using from the radius according to the equation

$$\text{Circumference} = 2 \times \text{pi} \times \text{Radius}$$

Combining these two equations, will lead to a new formula related to the speed of an object moving in *uniform circular motion* to the radius of the circle and the time to make one orbit:

$$\text{Average Speed} = [2 \times \pi \times R] / T$$

where R represents the radius of the circle and T represents the time period. Thus, a twofold increase in radius corresponds to a twofold increase in speed; a threefold increase in radius corresponds to a three-fold increase in speed; and so on. But this all is true for the 3D world. If this canon would work in 5D spacetime, we would assume that in order to exit the black hole we had to make a large radius motion inside of it, instead of trying to escape by a direct motion outward its horizon. This makes me believe, that the treatment

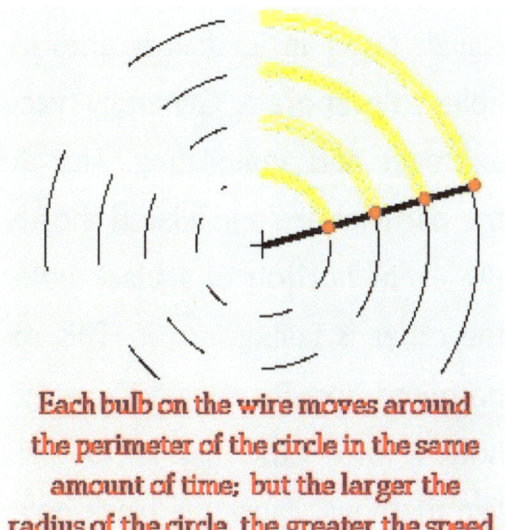

Each bulb on the wire moves around the perimeter of the circle in the same amount of time; but the larger the radius of the circle, the greater the speed.

of the evil in outer space varies of that on our planet. Instead of ignoring the evil, as we do here, one needs to play with it inside the black hole!" meowed the cub.

"Indeed. But several studies suggest, that the standard Einstein-Maxwell equations in 2+1 spacetime dimensions, with a negative cosmological constant, admit a black hole solution. The 2+1 black hole—characterized by mass, angular momentum, and charge, defined by flux integrals at infinity—is quite similar to its 3+1 counterpart. Anti−de Sitter space appears as a negative energy state separated by a mass gap from the continuous black hole spectrum. Evaluation of the partition function yields that the entropy is equal to twice the perimeter length of the horizon.[7]

"Negative energy vacuum?"

"Aha. One of the properties of negative-energy vacuum is that light actually travels faster in it, than it does in a normal vacuum. Negative energy also causes black holes to evaporate. Generally, vacuum energy is modeled as virtual particles popping into existence and annihilating. This doesn't violate any energy conservation laws as long as the particles are annihilated shortly afterwards. However, if two particles are produced at the event horizon of a black hole, one can be moving away from the black hole, while the other is falling into it. This means they won't be able to annihilate, so the particles both end up with negative energy. When the negative energy particle falls into the black hole, it lowers the mass of the black hole instead of adding to it, and over time particles like these will cause the black hole to evaporate completely. Because this theory was first suggested by *Stephen Hawking*, the particles given off by this effect are called *Hawking radiation,* Hawking's greatest scientific achievement - a part of his *Bing Bang* theory." [8]

"How do black holes evaporate?"

"This is a tough question. Back in the 1970's, Hawking came up with theoretical arguments showing that black holes are not really entirely black: due to quantum-mechanical effects they emit radiation. Suppose one of these vacuum fluctuations happens near the horizon of a black hole. It may happen that one of the two particles falls across the horizon, while the other one escapes. The one that escapes carries energy away from the black hole and may be detected by some observer far away. To that observer, it will look like the black hole has just emitted a particle. This process happens repeatedly, and the observer sees a continuous stream of radiation from the black hole. The energy that produces the radiation comes from the mass of the black hole. Consequently, the black hole gradually shrinks. It turns out that the rate of radiation increases as the mass decreases, so the black hole continues to radiate more and more intensely and to shrink more and more rapidly until it presumably vanishes entirely. A two-dimensional model has shown that using *fermion-boson cancellation* on the stress-energy tensor reduces the energy outflow to zero, while other *noncovariant techniques* give the Hawking result. A technique for replacing the collapse by boundary conditions on the past horizon retains the essential features of the collapse while eliminating some of the difficulties.[9] The spherically symmetric coupled scalar-gravitation Hamiltonian theorem gives the hope that someone can apply it to the problem of black-hole evaporation."

"How big is a black hole?"

"There are at least two different ways to describe its size!" replied the jaguar. "We can say how much mass it has, or we can say how much space it takes up. There is no limit in principle to *how much* mass a black hole can have. Any amount of mass at all can in principle be made to form a black hole if you compress it to a high enough density. It is suspected that most of the black holes that are actually out there were produced in the deaths of massive stars, and so we expect those black holes to weigh about as much as a massive star.

A typical mass for such a stellar black hole would be about 10 times the mass of the Sun, or about 10^{31} kilograms. 10^{31} means a 1 followed by 31 zeroes, or 10,000,000,000,000,000,000,000,000,000,000. Astronomers also suspect that many galaxies harbor extremely massive black holes at their centers. These are thought to weigh about a million times as much as the Sun, or 10^{36} kilograms.

The more massive a black hole is, the more space it takes up. In fact, theSchwarzschild radius (the radius of the horizon) and the mass are directly proportional to one another: if one black hole weighs ten times as much as another, its radius is ten times as large. A black hole with a mass equal to that of the Sun would have a radius of 3 kilometers. So a typical 10-solar-mass black hole would have a radius of 30 kilometers, and a million-solar-mass black hole at the center of a galaxy would have a radius of 3 million kilometers. Three million kilometers may sound like a lot, but it's actually not so big by astronomical standards. The Sun, for example, has a radius of about 700,000 kilometers, and so that supergiant black hole has a radius only about four times bigger than the Sun."

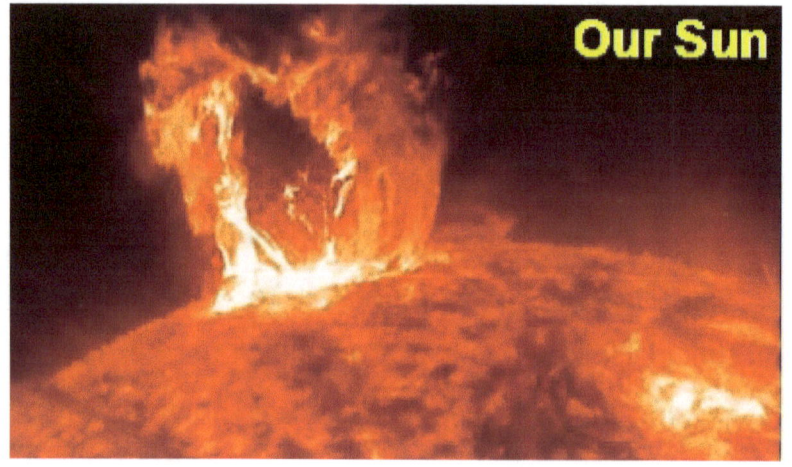

Our Sun

"What if the Sun became a

black hole? " mowed the cub.

"Well, first, let me assure you that our Sun has no intention of doing any such thing. Only stars that weigh

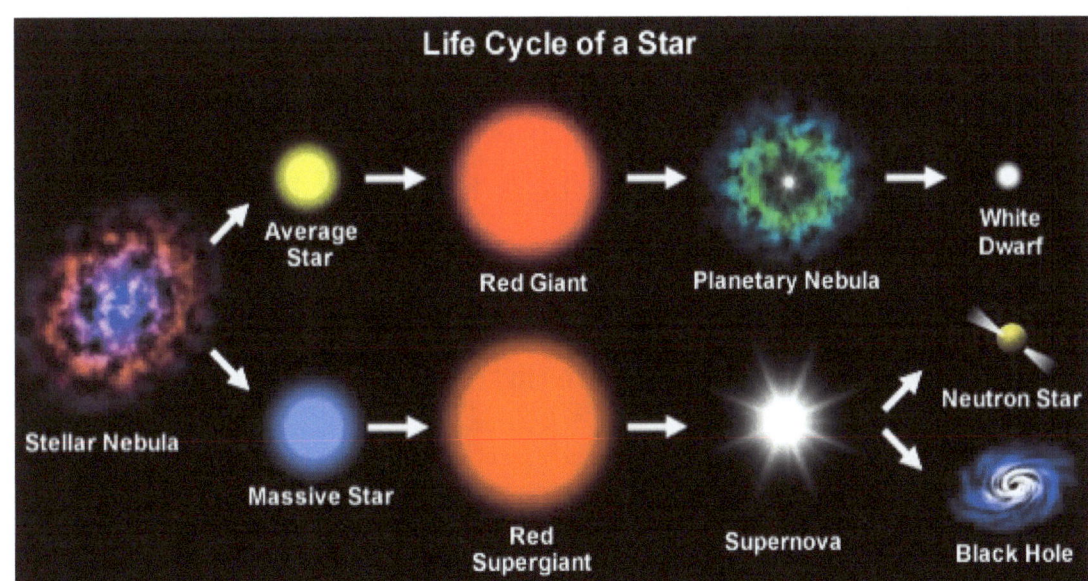

considerably more than the Sun, end their lives as black holes. Our sun is 4.5 billion years old, a mid-age average star. If you remember the formation of the stars and their life-cycle, our Sun is going to stay roughly the way it is for another 4.5 billion years or so. Then it will go through a brief phase as a red giant star, during which time it will expand to engulf the planets *Mercury* and *Venus*, and make life quite uncomfortable on Earth (oceans boiling, atmosphere escaping, etc). After that, the Sun will end its life by becoming a white dwarf. If I were you, I wouldn't buy any of those eight-billion-year government bonds..." roared the jaguar.

"Let's assume, our Sun did become a black hole for some reason. What would happen then?" purred the cub stretching out the tail.

"The main effect is that it would get very dark and very cold around here. The Earth and the other planets would not disappear into the black hole; they would keep on orbiting in exactly the same paths they follow right now."

"Why so?"

"Because the horizon of the black hole would be very small -- only about three kilometers. As wee discussed, as long as you stay well outside the horizon, a black hole's gravity is no stronger than that of any other object of the same mass."

"How many black holes are in our galaxy, *Milky Way?*" meowed the cub intimidated.

"There are infinite numbers of black holes in the space. Only our galaxy contains nearly 100 million black holes and this is when *Milky Way* is a mid-age spiral galaxy. Imagine how many black holes Andromeda would have, when *Andromeda* is twice our age, 9 billion years old!"

"How can we see the black hole, if the light itself is devoured by it? If we can't see the black hole, then how we prove its existence?" meowed the cub.

"Glad you finally asked! You can't see a black hole directly, since light can't get past the horizon. We rely on the indirect evidence that black holes exist. Suppose you have found a region of space where you think there might be a black hole. How can you check whether there is one or not? The first thing you'd like to do is to measure how much mass there is in that region. If you've found a large mass concentrated in a small volume, and if the mass is dark, then it's a good guess that there's a black hole there. There are two kinds of systems in which astronomers have found such compact, massive, dark objects: the centers of galaxies, and X-ray-emitting binary systems in our own Galaxy.

According to *Kormendy* and *Richstone,*[10] eight galaxies have been observed to contain such massive dark objects in their centers. The masses of the cores of these galaxies range from one million to several billion times the mass of the Sun. The mass is measured by observing the speed with which stars and gas orbit around the center of the galaxy: the faster the orbital speeds, the stronger the gravitational force required to hold the stars and gas in their orbits.

These massive dark objects in galactic centers are thought to be black holes for at least two reasons. First, it is hard to think of anything else they could be: they are too dense and dark to be stars or clusters of stars. Second, the only promising theory to explain the enigmatic objects known as quasars and active galaxies postulates that such galaxies have super-massive black holes at their cores. If this theory is correct, then a large fraction of galaxies - all the ones that are now or used to be active galaxies -- must have super-massive black holes at the center. Taken together, these arguments strongly suggest that the cores of these galaxies contain black holes, but they do not constitute absolute proof.

Several discoveries strongly support the presence of the black holes:

(1) First, a nearby active galaxy was found to have a '*water maser system'* near its

nucleus - a very powerful source of microwave radiation. Using the technique of very-long-baseline interferometry, the velocity distribution of the gas was mapped with a fine

resolution. In fact, the velocity within less than half a light-year of the center of the galaxy was measured. Thus, the massive object at the center of this galaxy is less than half a light-year in radius. It is hard to imagine anything other than a black hole that could have so much mass concentrated in such a small volume.[11]

(2) Another discovery provides even more compelling evidence securing the hypothesis of the black hole. X-ray astronomers have detected a spectral line from one galactic nucleus that indicates the presence of atoms

near the nucleus that are moving extremely fast (about 1/3 the speed of light). Furthermore, the radiation from these atoms has been red-shifted in just the manner one would expect for radiation coming from near the horizon of a

black hole. These observations would be very difficult to explain in any other way besides a black hole.[12]

(3) Next, there are much lighter, stellar-mass black holes in our galaxy, which are thought to form when a massive star ends its life in a supernova explosion. If such a stellar black hole were to be off somewhere by itself, we wouldn't have much hope of finding it. However, many stars come in binary systems -- pairs of stars in orbit around each other. If one of the stars in such a binary system becomes a black hole, we might be able to detect it. In particular, in some binary systems containing a compact object such as a black hole, the matter is sucked off of the other object forming an 'accretion disk' of stuff swirling into the black hole. The matter in the *accretion disk* gets very hot as it falls closer and closer to the black hole, and it emits copious amounts of radiation, mostly in the X-ray part of the spectrum. Many such *'X-ray binary systems'* are known, and some of them are thought to be likely black-hole candidates."

"Suppose you've found an X-ray binary system. How can you tell whether the unseen compact object is a black hole?" asked the cub.

"Well, one thing you'd certainly like to do is to estimate its mass. By measuring the orbital speed of visible star you can figure out the mass of the invisible companion. If the mass of the compact object is found to be too large, then there is no kind of object we know about that it could be other than a black hole. For instance, an ordinary star of that mass would be visible, or a stellar remnant such as a neutron star, would be unable to support itself against gravity, and would collapse to a black hole. The combination of such mass estimates and detailed studies of the radiation from the accretion disk can supply powerful circumstantial evidence that the object in question is indeed a black hole.

(4) Based on 26 galaxies observed by *Hubble Space Telescope* for stellar kinematics, *Gebhardt et al* have described a correlation between the mass Mbh of a galaxy's central black hole and the luminosity-weighted line-of-sight velocity dispersion σe within the half-light radius. The best-fit correlation they have cam up is:

$$Mbh = 1.2 \ (\pm 0.2) \times 10^8 \ M_\odot \ (\sigma e / 200 \ km \ s^{-1}) \ 3.75 \ (\pm 0.3)$$

over almost three orders of magnitude in Mbh; the scatter in Mbh at fixed σe is only 0.30 dex, and most of this is due to observational errors. The Mbh-σe relation is of interest not only for its strong predictive power but also because it implies that central black hole mass is constrained by and closely related to properties of the host galaxy's bulge.[13]

(5) Last not the least: a significant fraction of nearby galaxies show evidence of weak nuclear activity unrelated to normal stellar processes. High-resolution multi-wavelength observations indicate that the bulk of this activity derives from black hole accretion with a wide range of accretion rates. The low accretion rates that typify most low-luminosity active galactic nuclei induce significant modifications to their central engine. The broad-line region and obscuring torus disappear in some of the faintest sources, and the optically thick accretion disk transforms into a three-component structure consisting of an inner radiatively inefficient accretion flow, a truncated outer thin disk, and a jet or outflow. The local census of nuclear activity supports the notion

that most, perhaps all, bulges host a central supermassive black hole, although the existence of active nuclei in at least some late-type galaxies suggests that a classical bulge is not a prerequisite to seed a nuclear black hole."[14]

"I've heard, that there are also *white holes*," meowed the cub.

"What I admire, is that the concept of general relativity has a fascinating

mathematical property: *the symmetry in time!* Since a black hole is a region of space from which nothing can escape, the time-reversed version of a black hole is a region of space into which nothing can fall. In fact, just as a black hole can only suck things in, a white hole can only spit things out!"rambled the jaguar.

"What about the *vanilla holes?*"

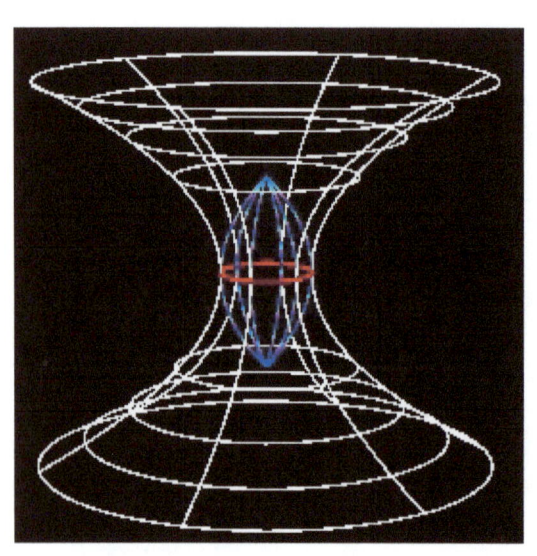

"Those are *wormholes,* or 'vanilla' black holes. If we consider black holes that rotate or have charge, things get more complicated. In particular, it is possible to fall into such a black hole and not hit the singularity. In effect, the interior of a charged or rotating black hole can 'join up' with a corresponding white hole in such a way that you can fall into the black hole and pop out of the white hole. This combination of black and white holes is called a *wormhole.*"[15]

"So, the white hole may be somewhere very far away from the black hole; indeed, it may even be in a different universe -- that is, a region of spacetime that aside from the *wormhole itself,* is completely disconnected from our own region. A conveniently-located *wormhole* would therefore provide a convenient and rapid way to travel very large distances, or even to travel to another universe. Maybe the exit to the *wormhole* would lie in the past, so that I could travel back in time by going through. All in all, sound pretty cool!" meowed the cub.

"But, before you formulate your hypothesis and defend a thesis, there are a couple of things you should know. First of all,

wormholes almost certainly do not exist. Just because something is a valid mathematical solution to the equations, doesn't mean that it actually exists in nature. In particular, black holes that form from the collapse of ordinary matter, do not form *wormholes*. If you fall into one of those, you're not going to pop out anywhere. You're going to hit a singularity, and that's all there is to it. Further, even if a wormhole were formed, it is thought that it would not be stable. Even the slightest perturbation would cause it to collapse. Finally, even if *wormholes* exist and are stable, they are quite unpleasant to travel through. Radiation that pours into the *wormhole* (from nearby stars, or the cosmic microwave background) gets blue-shifted to very high frequencies. As you try to pass through the *wormhole*, you will get fried by these X-rays and gamma rays."

"What is between all these holes? A vacuum?"

"Not necessarily. The standard concept suggests that the universe is a vacuum. However, it is established that ten thousand light years from earth, in a constellation Aquila, there are giant clouds of alcohol floating in space. Thus, the space has lungs!"

"But.... alcohol is an organic substance," meowed the cub. "Can it be true, that there is life outer our planet?"

"I would not comment on that one. What I can say, is that the cloud is 1000 times larger than the diameter of our solar system. It contains enough ethyl alcohol to fill 400 trillion-trillion pints of beer. To down that much alcohol, every person on earth would have to drink 300,000 pints each day—for one billion years. Further, in a distance of 58 quadrillion miles away from the earth, there is a cloud- cocktail of 32 compounds: among them *carbon monoxide, hydrogen cyanide,* and *ammonia.* Lastly, our galaxy has a second intergalactic liquor cabinet in the *Sagittarius B2 Cloud,* which holds 10 billion-billion-billion liters of cosmic hooch. The cloud holds mostly methanol (CH_4O). Similarly, near the center of the *Milky Way,* a cloudy bridge of methanol surrounds a stellar nursery. Now, if you're wondering what these space spirits may taste or smell like, the cloud at Sagittarius B2 contains *ethyl formate*, reportedly smelling like a raspberry rum! *Ethyl formate* ($C_3H_6O_2$) is an ester formed when *ethanol* (C_2H_6O) reacts with *formic acid* (CH_2O_2):

$$C_3H_6O_2 \ = \ C_2H_6O \ + \ CH_2O_2 \ - \ H_2O."$$

"Is there an interpretation?"

"It wasn't spilled after some *Martian keg party*. As new stars heat up—formed as clouds of gas and dust collapse—ethyl alcohol can attach to specks of floating dust. As the dust moves toward the budding star, the alcohol heats, separates, and turns to gas. For astronomers, these alcohol clouds can be a clue into how our biggest stars form. And yes, the alcohol is an organic compound. Therefore, these drunken clouds may help us better understand how life might arise elsewhere in the cosmos!"

"In order the alcohol reach the free space, biological organisms are needed to produce some kind of sugar. The alcohol

molecule might be formed by the random interactions of floating atoms in space, but not billions of gallons in large chunks," argued the cub.

"Formerly, the scientists considered that some molecules might gather on bits of dust floating in the vacuum of space. The surface of the dust might let these molecules interact and form alcohol. Fast-moving molecules might then blow the alcohol off the dust, leaving gallons of it in space. However, there wasn't any conceivable way to peel the alcohol molecule off the dust without destroying the structure of the molecule in space. Currently, the scientists suggest that ice could form on the dust, trapping the alcohol. As the ice melts and evaporates when the dust bit drifts near new star clusters, the alcohol is gently freed without getting destroyed."

"What a journey!" exclaimed the cub. "Now, once we learned how to escape the black holes, and how to stay sober, are you thinking what I'm thinking?"

"You mean the *Mission Jupiter-P*?"

"Aha. The garlic odor of Jupiter always captivated my curiosity. My mission is to explore the compound *phosphine'* in that giant planet. Accordingly, I name my mission *'Jupiter-P.'* Phosphine (PH_3) is a toxic compound with garlic odor. Unlike our planet Earth, Jupiter's atmosphere does not have oxygen and therefore, there is no life in this outer planet. *Phosphates* (PO_4) or *phosphids* (K_3P, Na_3P, or Ni_5P_2) - the basis of the organic matter, are typical only to the planet Earth. Jupiter does not have free phosphorus (P), and carries only traces of *phosphine* (PH_3), a toxic compound with the garlic odor. My task is to found out whether the Jupiter's phosphorus can be oxygenated if a sample of it would be brought onto Earth. Having the oxygenated compound, we would then return it to Jupiter to explore, whether any regenerative reaction on the surface of Jupiter would release a free oxygen in the atmosphere, and if I would be the one who would spring life on Jupiter!"

"Well said! However, in technical terms it is *uneasy* to reach the planet Jupiter," reminded father jaguar. "Flights from Earth to other planets of our Solar System have a high energy cost. In astrodynamics this energy expenditure is defined by the net change in the

spacecraft's velocity, or *delta-v.* The energy needed to reach Jupiter from the Earth's orbit requires a delta-v of about 9 km/s - compared to the 9.0–9.5km/s needed to reach the Earth's orbit. "

"Scientists have developed *ion thrusters,* capable of a *delta-v* of more than *10 kilometers/s.* This is more than enough do reach Jupiter without gravity assist," mewled the cub. "Currently, the distance between the planets Earth and Jupiter is 867-million kilometers. Therefore, with a velocity of 10 km/s it will take me 86,700,000 seconds, or 1,445,000 minutes, or 24,083.33 hours, or 1,003.4722 days, or 2.75 years to reach Jupiter."

"The day you land on Jupiter, you'll be a grown-up jaguar!" sighed father jaguar.

"... if using the time estimate applicable for our planet," meowed the cub. "Time is different on Jupiter. A day on Jupiter lasts 9.92496 hours. Thus, a day spent on Earth roughly equals 2.4 days spent on Jupiter. That assuredly would make me older than you think..."

"Not that easy!" thundered the jaguar. "Determining the length of a day on Jupiter is very difficult, because, unlike the terrestrial planets, it does not have surface features that could be used to determine its rotational speed. Scientists only judge the planet's rotational speed. An early attempt was to do some storm watching, as Jupiter is constantly buffeted by atmospheric storms. Locating the center of a storm would help understand the length of a day. The problem was that the storms on Jupiter are too fast , making them an inaccurate source of information. Scientist were finally able to use radio emissions from Jupiter's magnetic field to calculate the planet's rotational period or speed."

"I agree. Jupiter is almost a solar system unto itself. This planet resembles a failed star - lacking the mass needed to ignite fusion. Due to its gravity, it has 50 confirmed

moons and at least 14 provisional moons. The four largest moons (*Galilean Io, Europa, Ganymede*, and *Callisto*) are all scientifically intriguing. *Io* is a volcanic nightmare. *Europa* is covered in ice and may have oceans of slushy ice underneath. *Ganymede* is the largest moon in the solar system, bigger than *Mercury*, and is the only moon known to have an internally generated magnetic field like Earth's. *Callisto* is interesting because its surface is thought to be very ancient; perhaps original material from the birth of the solar system."

"Now, once you know the turbulent and weird 'life' of Jupiter, have you decided what type of engines and instruments you would need to complete your mission P?"

"For sampling *phosphine* on Jupiter's surface, as the minimum I would need a vacuum chamber, a sterile vacuum gadget spoon, a piece of radiation shielding black paper for wrapping my tube - as the Jupiter's blood-red radiation belts may damage my sample, good connection with Earth, eye-shield for the infra and ultra, a family album to cool off my nostalgic attacks, a mirror, a toothbrush, a thermometer and barometer, a compass and mechanical watch, an android, pills for constipation, and of course enough supply of food and oxygen... I think that's enough for my juvenile spaceship, as my task is too simple. Sunset arrives. Twilight announces its reign. It's time to dispatch. Four, three, two, one.... bingo!" the cub dressed the whiskers, comely put on the spacesuit, and pushed the button.

REFERENCES:

1. Klein O (1926). Quantentheorie und fünfdimensionale Relativitätstheorie. *Zeitschrift für Physik;* A 37(12): 895–906

2. Thiry MY (1948). Cosmological solutions. *Compt. Rend. Acad. Science of Paris;* 226: 216–218

3. Ludwig G, Müller C (1948). Ein Modell des Kosmos und der Sternentstehung. *Ann. Phys. Leipzig;* 2(6): 76–84.

4. Scherrer W (1949). Über den einfluss des bauordnungsrechts auf die qualität. *Helv. Phys. Acta;* 22: 537–551

5. Dehnen H, Obregón O (1971). Exact cosmological solutions in Brans and Dicke's scalar-tensor theory. *Astrophysics and Space Science;* 14(2):454-459

6. Williams LL (2012). Physics of the electromagnetic control of spacetime and gravity. Proceedings of the *48th AIAA Joint Propulsion Conference;* AIAA 2012-3916

7. Bañados M, Teitelboim C, Zanelli J (1992). Black hole in three-dimensional spacetime. *Physics Revolution;* 69: 1849

8. Bunn T (1995). Black holes. *Berkeley Cosmology Lectures.*

9. Unruh WG (1976). Notes on black-hole evaporation. Physics Revolution; 14:870

10. Kormendy J, Richstone DO (1995). Inward Bound: The search for supermassive black holes in Galaxy nuclei. *Annual Review of Astronomy and Astrophysics;* 33: 581

11. Miyoshi M, Moran J, Herrnstein J, et al (1995). Evidence for a black hole from high rotation velocities in a sub-parsec region of NGC 4258. *Nature*; 373:127-129

12. Tanaka Y, Nandra K, Fabian AC, et al (1995). Gravitationally redshifted emission implying an accretion disk and massive black hole in the active galaxy MCG 6 30 15. *Nature;* 659-661

13. Gebhardt K, Bender R, Bower G, et al (2000). A relationship between nuclear black hole mass and galaxy velocity dispersion. *The Astrophysical Journal Letters*; 539:1

14. Ho LC (2008). Nuclear activity in nearby galaxies. *Annual Review of Astronomy and Astrophysics;* 46: 475-539

15. Macher J, Parentani R (2009). Black/white hole radiation from dispersive theories. *Physical Review D*; 79(12):124008

www.ingramcontent.com/pod-product-compliance
Lightning Source LLC
Chambersburg PA
CBHW050358180526
45159CB00005B/2062